Canadian Railway Scenes

Nº 4

COMPILED BY
Adolf Hungry Wolf

CONTENTS

'The Canadian' — First Days, Last Days 2
Diesels on the Hight Iron — 1956 — Eric Grubb 22
1952 CPR Caboose Ride through the Rockies 24
Canadian Cabooses: A Farewell Photo Tribute 34
Mixed Train Caboose Ride — Omer Lavallée 50
CN at Night 52
Ontario Steam Wayfreight — James A. Brown 54
Final Steam Rebuilds at CN Stratford — James A. Brown 56
Great Canadian Steam Railway Excursion 60

(Back cover) C.P.R. Cabooses, Minnedosa Hill, on the Canadian Prairies of Manitoba; Summer 1958.
Photo by Mavis Stuckey.
(Front cover) Doubleheaded Steam Today: British Columbia's ex-CPR 3716 and 2860 lead the Great Canadian Steam Railway Excursion south over BC Rail tracks, Autumn 1988.
Photo by Adolf & Okan Hungry Wolf.
(Inside Front Cover) Back when the crews lived with their engines and cabooses. CPR No. 307 with caboose 3685 at Canmore Alberta; c. 1907.
CP Rail Corporate Archives
(Inside back cover) Downtown Toronto in 1949 — CN Spadina with a full house.
Adolf Hungry Wolf Collection.

Paste-up and Design by Adolf Hungry Wolf
Typesetting by Rocky Mountain Printers Ltd.
Printed in U.S.A. by McNaughton & Gunn
First Printing — Spring 1991

ISBN 0-920698-25-5

C.P.R. Train #2, "The Canadian," arriving at Kenora, Ont. c. 1956.

Keith Dickens Photo
Paterson-George Collection

Nicholas Morant-CP Archives.

CP Archives

'THE CANADIAN'
FIRST DAYS — LAST DAYS
1955 — 1990
A Photo Tribute to Canada's Greatest Train

[Opposite Page] First Days of 'The Canadian,' as eastbound Train No. 2 pulls into the famous tourist destination of Banff, Alberta. The crowd illustrates summer dress before shorts, in 1955, while the latest styles from Greyhound and VW wait for arriving passengers in the bus parking area. Maroon, grey and yellow FP7A No. 1426 (GMD 9/52; later CPR 4070, then MUCTC 1300) leads a B-unit and a Geep, trailed by 13 cars. At this time Pullman service was still available to Banff, as evidenced by the heavyweight sleeper on the next track. Trains also brought most fruit and vegetables to Banff in the reefers seen at left. Today, nothing from this scene remains in Banff except for the tracks — and of course the trees and mountains.

[Right] Last Days of the 'Canadian' (only CP called it '*The* Canadian') saw engines like F40PH-2 No. 6408 (GMD 1986) leading as motive power, but the consist and station stops along the route remained practically the same. This scene is in front of the Banff station.

Adolf Hungry Wolf

Adolf Hungry Wolf

[Left] When VIA Rail took over 'The Canadian,' former CN cars in VIA's blue and yellow became mixed in with the stainless steel consists, as seen on both sides of this Skyline Dome car, at Banff in 1985.

Both, CP Corporate Archives

[Left] 'The Canadian' visits southeastern B.C. Prior to the April 1955 inauguration of the train, two of its stainless steel cars were sent on a tour of the CP system to give Canadians a preview. Here we see the pair parked in front of the wooden two-story station at Nelson, B.C. While workmen apply the finishing touches, police and officials prepare to handle the crowds. Across the yard tracks sits Nelson's new diesel shop, another sign of changing railroad ways, although some steam engines still operated into Nelson at this time. An A-B set of CLC-built "C-liners" stands in the doorway of the shop. How many of the hoods and grills parked by the station can you identify?

[Below] Triple Rare Photo! If there are few photos showing steam near equipment of 'The Canadian,' there are fewer still of such equipment actually hauled by steam. However, as the stainless steel cars arrived from builder Budd, prior to April 1955, they were put into service on various trains, both to break them in and get them to work, and also to get added publicity. This September 1954 scene shows steam-powered Train No. 39 approaching Mt. Orford, Quebec with a brand new dome car on the rear. And that's perhaps the rarest part of all in this photo, that the car is FUNDY PARK, of which very few photos exist. Out of the original 18 dome lounge-observation cars built, only "Fundy Park" is missing, it having been wrecked during 1957 in a rear-end collision at Gull Lake, Sask.

[Right] Perhaps the best-known scene of 'The Canadian' is this shot taken by the photographer at what is now known as Morant's Curve, named in his honor, along the Bow River, in the midst of majestic Banff National Park. Among other things, this shot was turned into a painting that adorned the cover of CP's 1955 public timetable. The 16-car westbound is trailed by dome car "Tremblant Park." As a personal note, this print was given to me by Nick Morant while we shared breakfast at his home in Banff, after which he drove me to the station, where I boarded VIA's 'Canadian' for another trip east. Our dome car was "Tremblant Park."

[Left] Although CP operated steam for another five years after the first cars arrived for 'The Canadian,' photos showing the two together are rare. This scene was cropped from a larger publicity photo, though the presence of steam made it useless for the company's campaign to project a modernizing image.

Nicholas Morant, CP Rail

RNE/AHW Collection

[Left] N.R. "Buck"Crump greets the train's sleeping car porters, back in an era when this was about the only job non-whites could obtain aboard the train.

[Left Below] Engineer and Conductor compare watches in the company of CP Vice-President Crump (left) and Montreal Mayor Jean Drapeau. Crump became CP's President shortly after this, in part for his successful inauguration of 'The Canadian,' although he later regretted making this notable contribution to Canadian travel and history.

[Below] All youngsters on the first train were presented with HO models of its FP7A locomotive, such as these two aboard the rear dome car.

[Upper right] CPR Chairman W. A. Mather "pulls the switch that signals the start of 'The Canadian' as Mayor Jean Drapeau of Montreal looks on."

[Lower right] Mayor Drapeau and Vice-President Crump check out the engineer's side of the first train's FP7A locomotive.

[Below] Smiling CPR Sleeping Car Conductor accepts the first ticket for a private space aboard the new train, which will pull out promptly in another half hour. Send-off music was provided by the Victoria Rifles Regiment Band, which had also played upon departure of the very first transcontinental train.

THE FIRST TRAIN

April 24, 1955

" 'The Canadian', the Company's new all stainless steel scenic dome train left Montreal and Toronto amid pageantry and fanfare as it set out to establish a new transcontinental record for railway passenger service to Vancouver. . ." So said an article in the CP "Spanner" the following month, further explaining that the new train "travels the 2,881 miles from Montreal to Vancouver in 71 hours and 10 minutes, as compared with the former transcontinental time of 87 hours, 10 minutes." Leaving Montreal at one p.m. on Sunday, it arrived in Vancouver at 9:10 a.m. on Wednesday, April 27th. Its other notable distinctions included, "a) the longest continuous run of a diesel locomotive; b) the world's longest Scenic-dome ride (up to the time of its 1990 demise!); c) Canada's first and only fully streamlined, stainless steel, mechanically air-conditioned train and d) the first scenic dome and skyline coach-dome luxury cars to operate on Canadian rails." On these two pages are CPR publicity photos taken just before the first train's departure.

Six photos, CP Corporate Archives

Nicholas Morant, Photo/Canadian Pacific Corporate Archives

[Above] 'The Canadian' in its first days, along the shores of Gap Lake in Banff National Park, a highlight in its cross-country journey. In back are Mt. Lougheed (10,190') and Windtower (8,850'). Engine 1417 (GMD PF7A, blt. 1952) was destroyed by fire in 1971 after falling into a turntable pit in Vancouver.

[This page] Stoney Indians inspect the newest conveyance to enter their land, as dome car "Banff Park" makes its first visit to the place it was named for, located in the Stoney's traditional domain. Dressed like other Canadians most of the time, these tribal leaders donned feathers and buckskins to pose for CPR publicity photos. Unfortunately, their scenic reserve just outside the National park has been among the poorest in the country, so that most Stoneys never got to ride the sleek domeliner that passed nearby their cabins every day for 35 years.

Both, CP Corporate Archives.

Philip R. Hastings Photo

'The Canadian' at its three largest eastern destinations: [Above] Windsor Station in Montreal was home base for the train from the time it started in 1955 until VIA Rail took over in 1978 and moved it to the nearby CN station. Dome car "Sibley Park" was on the rear of Train No. 1 at Windsor Station in this view of October 1978. The train is ready to leave from Track 3, while on Track 4 sits Amtrak No. 159, a Rohr-built Turbo engine on Train No. 68, the 'Adirondack.' On track 5, CPR Dayliner RDC No. 9058 has just come in from Ottawa as Train No. 132.

[Opposite top] The Toronto-Sudbury section of 'The Canadian' was for many years assigned to locomotives which were not in the transcontinental passenger pool. These included MLW-built cab units and road switchers, such as FPA-2 No. 4082, seen here with GMD-built F7B No. 1915 on Train No. 11, departing West Toronto; Sept. 15, 1959.

[Below] Train No. 1, the Montreal section of 'The Canadian,' is seen arriving at Ottawa Union Station, May 3, 1963. The coach yard is moderately full, the switchstands still have lamps and the station is still downtown in the nation's capitol. All this has changed. Note that the train has two conventional heavyweight tourist sleepers behind the baggage cars, instead of the usual U-class stainless steel cars.

James A. Brown Photo

James A. Brown Photo

James A. Brown Photo

[Opposite] One of many dramatic locations along the transcontinental route of 'The Canadian' was this crossing of the Little Pic River, near Middleton, Ont., between White River and Schreiber. The scene shows westbound No. 1 on Sept. 10, 1965.

[Right] CP No. 1420 (FP7A) and No. 1910 (F9B) power 'The Canadian' — Toronto section, train 12 — southbound over the high bridge at Parry Sound, Ont., Aug. 14, 1961.

[Below] FP7A Nos. 1418 and 1407 speed 'The Canadian' westbound across part of the Ottawa River at Dorion, Quebec in 1971. Note CP president N.R. Crump's business car 'Laurentian' in the consist.

James A. Brown Photo

Bob Sandusky Photo

Adolf & Okan Hungry Wolf

[Left] Fireman's view from lead locomotive of the eastbound 'Canadian' looks out at snow-covered countryside in the rugged Selkirk Mountains, near Glacier station and the Connaught Tunnel. A westbound ballast train approaches behind a pair of CP Rail SD40-2's, on this March morning in 1986.

[Below] The photographer believes this was the first meet of eastbound and westbound sections of 'The Canadian' in Banff National Park, after the train began service in 1955. Dome car 'Yoho Park' has just left its namesake, while FP7A No. 1419 leads its train there.

Nicholas Morant Photo/Canadian Pacific Corporate Archives

[Right] Dome car view of steam in the early morning light. On its cross-country tour in 1978, British Columbia's ex-CPR 'Royal Hudson' No. 2860 'took the hole' at Stralak, Ont., 11.5 miles west of Cartier, to allow the eastbound 'Canadian' to pass. It is seen from upstairs in car 'Sibley Park.'

[Below] CP No. 1431 slows 'The Canadian' westbound at Canmore, Alberta on Nov. 6, 1960 to meet its eastbound counterpart just outside Banff National Park.

James A. Brown Photo

Bob Sandusky Photo

VIA RAIL'S ORIGINAL 'CANADIAN' IN ITS FINAL DAYS

In March of 1989 the Canadian government announced massive cutbacks in its federally-operated passenger train services. Still, no one could believe that the major fatality of this program would be **the** 'Canadian,' one of the world's last great train rides, whose seats were often sold out months in advance right up to the very end. On these pages are scenes from the final days of North America's last daily-scheduled domeliner.

[Left] Two weeks to go before CP Rail's mainline through the Canadian Rockies would lose two of its classic institutions — cabooses and dome cars — both seen here at the crew-change point of Field, B.C.

[Below] Dome car 'Yoho Park' at near sixty miles per hour, at the tail end of the eastbound 'Canadian,' Train 2, just out of Lake Louise, Alberta.

Both pages, Adolf and Okan Hungry Wolf Photos.

[Above] Train 2 at Morant's Curve, along the Bow River, rolling through Banff National Park. A CP Rail SD40-2 leads the consist of original 'Canadian' stainless-steel cars.

[Above right] Mountain meet between Trains 1 and 2 at Canmore, Alberta, with part of the fabled Three Sisters on the far right.

[Bottom] Built in 1909, this log station was already old when 'The Canadian' first began to stop here, 35 years ago. Although passenger service to here has now ended with the train's demise, the station has been given historic landmark status and will be preserved.

LAST TRIP

The last 'Canadian' departed from Montreal on a cold, grey Sunday morning, January 14, 1990, setting out on its historic transcontinental journey with a full house of journalists and nostalgia-seekers, greeted by protestors and placards at every stop. Many places along its route would be without passenger train services for the first time in over one hundred years, after its passing.

{Below] Media crowds and masses of protestors accompanied the final 'Canadian' clear across the country, as seen here at the nation's capitol in Ottawa, making this an unusual train trip in more ways than one.

[Right] Dining car crew aboard the last train looks out at the protesting crowds during the train's stop in Regina, as seen across the platform of a CP Rail caboose, whose days are also ending.

[Opposite, top] CP's stately Regina station will no longer see passenger service after this last train departs.

[Below] Rear dome of the 'Canadian' was busy day and night during this last journey, as media folks and passengers conversed on the merits of the train and the results of its demise. Ironically, the car was 'Banff Park,' first sent on a tour of Canada when brand-new in 1954 to promote the upcoming streamlined train.

[Far right] Medicine Hat was another stately CP station to lose all passenger services after departure of this, the last train.

Both pages, Adolf and Okan Hungry Wolf Photos

[Far left] Looking back at Medicine Hat for the last time in the icy dusk of January 1990, a yard switcher's headlight bids passenger train service farewell. By the opposite track stands another remnant from former times, warning crews on car roofs of the upcoming bridge. This was outdated years ago, when roofwalks were removed from freight cars.

[Upper left] Six hours late and without rights in CP's new timetable, the last 'Canadian' makes a rare daylight run down the Fraser Canyon as an "extra" train. While passengers enjoy the final hours, a CN freight passes across the canyon in the opposite direction.

[Lower left] Further down the Fraser Canyon, we meet yellow and silver Sperry Rail Service car no. 138, which once hauled passengers for the Rock Island lines.

Both pages, Adolf and Okan Hungry Wolf Photos.

VIA Rail Canada

[Above right] The late afternoon sun highlights this final meeting of CP's former star train and a CP caboose awaiting its last assignments, as the two pass in CP's Coquitlam yard.

[Below right] Just before sunset the last 'Canadian' makes its final entry into Vancouver, bringing to a close more than 100 years of transcontinental passenger service over Canadian Pacific's mainline. Only time will tell if this move results in better travel efficiency, or in the permanent loss of a valuable travel alternative.

DIESEL ON THE HIGH IRON

By Eric A. Grubb

(Excerpts from a story in the C.P. *Spanner,* March, 1956)

Railroaders can be a sentimental lot. To hear them tell it — when Montreal is Dieselizing a road job for them — there was never anything like reciprocating steam. They do, that is, until the new Diesels come to town!

Once they've seen them pull a train, and taken a look into the cab with its foam-rubber seats and wide, safety-glass windshield built so they can see the whole track ahead, with a set of controls they can operate with their fingers, and heaters and defrosters that will take them through a blizzard in shirt sleeves — then you don't hear so much about steam. Instead, you'll hear considerable talk among the veterans as to how much seniority they've got and who will get the Diesel and what runs they will take over.

Engineer Charlie Klein is a C.P.R. man who railroaded nearly 40 years with steam. Now it's all Diesel and that's okay with Klein. He runs transcontinental No. 2, the new ''Canadian'' streamliner, and he's waiting this afternoon at Field, B.C., to take the train over the 136 mile Rocky Mountain run to Calgary.

Smoke and steam are gone forever from the mountains but it's still railroading and there is plenty of drama and skill in the job. Climb aboard the head-end and watch Klein handle 4,500 horsepower across the shoulder of the Rockies. . .

''Highball!'' The time is 2:20 and Conductor Taylor's starting signal chirps twice in the cab. Klein notches out the throttle, the 1424 rumbles with power, and generators, motors and driving wheels lean into the job of rolling the streamliner out of town.

Gently, smoothly, we pick up speed through the yards. Ahead, the mainline signal flashes green. ''Clear Block!'' calls the fireman. ''Clear!'' repeats Klein. Then, as the tracks turn directly toward the mountains, Klein peels the throttle back and all the Rockies echo with the roar of Diesel power heading for Kicking Horse Pass in high gear.

We're moving up the famous ''Big Hill'' — 14 miles of twisting, looping pull up 1,260 feet from Field to the Continental Divide at Stephen. In the early days it took four steam engines to haul 700 tons to the top of this grade. The 1424 (A-unit with two B-units) spurns assistance with 1,000 tons in tow. What's more, she'll go over the summit in five minutes less time than the big '5900-class' steam locomotives — the most powerful in the British Empire — took with the ''Dominion'' just a few years back.

It's a big contrast to a steam job, but Klein smiles when you mention ''personality.'' ''Each steamer had its own quirks and you had to know how to nurse and coax them along. It was always a challenge to run them,'' he admits.

This 1424 cost about $100 per unit of horsepower. It's factory geared to cruise at 89 miles an hour with ''The Canadian'' and, what's more, it makes only 29 stops on the entire 2,881 mile haul, most of which are just for a change of crew. In the days of steam, a trans-continental passenger run required four locomotives which had to stop 65 times for fuel and water.

Although dubbed ''Diesel,'' the 1424 is basically an electric locomotive carrying its own generating plant. Prime mover in each of the three power sections is a 16-cylinder, two-cycle V-type Diesel engine. Each of these 1,500 horsepower units spins a generator which feeds direct current to traction motors down below, on the driving wheels, and that's what pulls the train.

As the long silver train coils up into the mountains, Klein keeps 600 volts or better reaching the traction motors. The throttle is in the last notch, our Diesels are making an awesome haul, but we're taking it steadily at 21 miles an hour. . .

Up near Cathedral, four miles out (of Field), we come to the daring Spiral Tunnel climb. From this spot, the tracks originally marched straight up to the gates of Kicking Horse Pass on a maximum grade of 237.6 feet to the mile. . . In 1909 railroad experts cut twin tunnels, figure-8 style, into opposite walls of the narrow valley and halved the grade in much the same way that an automobile road reduces a steep slope by hairpin turns. The old right-of-way has become part of the Trans-Canada Highway through Yoho National Park.

At 2:34 Klein flips his headlight switch and ''The Canadian'' burrows like a shiny gopher into the 2,922 foot bore beneath Mt. Ogden. For 90 seconds the tunnel walls and ceiling beat back the throbbing roar of the motors in deafening cadence as we corkscrew 240 degrees around the compass to gain 50 feet of altitude deep inside the mountain. Then the 1424 blares out into sunlight and we're headed west — not east — toward Cathedral Mountain and the mouth of Spiral Tunnel No. 2. You're nicely oriented to this change in direction when the rails dive into the mouth of the second tunnel for 3,255 feet, twirl the 15-car train around 250 degrees of a circle and set it back on an easterly course again.

When the 1424 noses out of Tunnel No. 2 you're high on the side of Cathedral Mountain where you can look into the depths of Yoho Valley and see the railroad doubling up the slope below. . . At Partridge siding (Fireman) McKenzie points out the remains of a 1925 rock slide which buried Yoho station on the level below and carried away portions of the track. The siding is named for the fireman of a westbound train who warned the occupants of the stationhouse and prevented any lives from being lost.

Suddenly Klein blasts away on the horn. Three Elk stand squarely between the rails about 50 yards ahead. For a moment the animals stare indifferently and ten bound gracefully into the forest.

It's 2:58 when the streamliner drones over the summit of Kicking Horse Pass and glides past a small brook which splits its waters between Alberta and British Columbia. . . A monument placed in the angle formed by the dividing waters commemorates the exploration work of Sir James Hector who discovered the Pass in 1858.

We drop sharply away from the 5,332 foot altitude at Stephen on a 1.8 per cent grade. There isn't a brake shoe against a wheel, but the heavy train rolls smoothly downward at a steady 30 miles an hour. Dynamic brakes — not friction braking — has the train under perfect control. Klein simply set his speed at the top of the hill, then flopped a switch and the traction motors became generators. The resistance holds back the train; power produced downhill is dissipated through grids in the locomotive roof.

The Diesel dynamic brake has added a great new dimension to Rocky Mountain railroading. "We seldom have to use our air-brakes anymore, except to stop," says Klein. Trains now get down steep grades without delay and, because the traction motors control the speed, the enormous cost of worn wheels and worn brake shoes is eliminated too.

At Lake Louise "The Canadian" pauses only momentarily to load vacationers. Klein gets the 'highball' at 3:10 and we rapidly pick up speed, the Diesels drumming behind us.

The 1424's gently, gliding starts are due partly to Klein's skill and partly to the 'load generator' which is arranged so that the amperage cannot mount too rapidly and hence, when taking off, the locomotive doesn't get an overdose of power. Most Diesels are set so that maximum power is attained in about 20 seconds and that is ample time for a smooth start.

We're rolling along briskly when No. 2 goes by Castle Mountain. "Yellow board!" calls McKenzie. Klein echoes "yellow!" and ratchets down the power.

"That's No. 1 heading into the passing track at Massive," McKenzie tells you. "He'll be in the clear by the time we get down there." No. 1 is our westbound counterpart. On single-track operation, railroad etiquette gives eastbound trains precedence over those bound west and so we hold the mainline.

No. 2 receives a green signal into Massive and there's the Montreal-Vancouver "Canadian" waiting on the siding behind three unit Diesel No. 1407. Porters and train crew members wave from the vestibules of silver sleepers and dome cars as we reel by.

Ahead, a bit later, McKenzie points out a pair of moose, oblivious of "The Canadian" drinking in the hunter-proof safety of the National Park. Then, at 3:55, the train pulls into Banff.

While passengers climb aboard and baggage is loaded, McKenzie takes a quick swing around the engine rooms. Through the long, narrow aisles of the three power sections, up one side and down the other. He checks the lubricating oil and cooling systems.

You stand in the cab doorway watching an elderly gentleman and a small boy with a box camera. The boy is photographing your Diesel.

"When we had the big '5900-class' steam engines there was always someone up there," Klein reflects.

"Highball!" Again Klein latches out the throttle and No. 2 glides out of Banff on the dot of 4 p.m. There are no more scheduled stops until Calgary where a new crew takes over for the lap to Medicine Hat."

G.C. Corey /RPS / AHW Colln.

1952 CABOOSE RIDE THROUGH THE ROCKIES

A Photo Journey

It was still dark one Fall morning in 1952 when C.P.R. 4th Class Extra #83 pulled out of Calgary westbound behind a new A—B—A diesel set led by No. 4036A. Behind the engines came 72 cars, trailed by wooden caboose No. 437023. This train was bound for glory, its day-long mountain run recorded for posterity by three very busy members of the Rail Photo Service, an ambitious group of American photographers based in the East.

The trip was led by H. W. Pontin, himself a venerable steam engineer on mainline trains in the U.S. Respect for their work was such that most railroads willingly lent support, including the C.P.R. in providing them with this classic caboose ride.

A major theme in many Rail Photo Service journeys was to show the transition from steam to diesel power. For that purpose alone, this 1952 trip through the Canadian Rockies was ideal. Diesels were quite common already, but steam still had a couple more years left, as well. Trains were still comparatively short, which meant a lot more of them. The following pages show only the highlights from that day's journey.

(All photos reproduced here were taken by H. W. Pontin and Warren G. Fancher from CPR caboose 437023. Negatives and prints from the vast holdings of the Rail Photo Service have become widely dispersed. Those shown here are from the Good Medicine Foundation, courtesy of the Railway Negative Exchange and Warren E. Miller)

[Above left] Conductor and brakeman stand on the back platform of CPR caboose No. 437023 (blt. c. 1943), from which the photos on these next pages were taken.

[Below] The first photo taken on this journey was of a three-way meet at Exshaw, Alberta by the first light of day. 'C-Liner' No. 4054 (CLC 8-52) was in its first month of service, though it soon left the mainline for work in the Crowsnest Pass region. It was wrecked in 1967 after hitting a rock slide near Nelson, B.C. Across the tracks we see part of GP7 No. 8410 (GMD 3-52), which was also less than a year old.

[Opposite, top and bottom] Lighting is better by the time we roll into the yard at Banff, where elderly 2-8-2 No.. 5248 (MLW 10-12; scr. 8-57) is doing some light switching. A steam engine was regularly assigned here in part to provide heating for sleepers and private cars parked on the left.

[Top right]GMD-built engines meet GM-built car, as a Pontiac passes on the adjacent gravel highway through Banff Park.

5248

4028

[Opposite, bottom and upper left] Our westbound train passes a similar one stopped in the yard at Lake Louise. Its power is an A-B set led by No. 4028 (GMD 9-50), CP's first FP7A, destroyed by fire in 1965. Trailing the train is an all-steel end-cupola caboose, whose trainman is on the ground with the coffee pot.

[Top right] At the road crossing just outside of Lake Louise we meet a pair of vintage Plymouths, station wagon and coupe, waiting for our train to pass.

[This page] Stephen, on top of the Continental Divide, highest point on the CPR mainline (nearly a mile above sea level, named after George Stephen, the railway's first president). Although isolated and often snowbound, the wood-frame station shows signs of occupancy by a family, including a homemade kids' tractor by the back door and well-tended flower boxes. Behind the station runs a unique little stream whose waters naturally divide, part flowing east to the Atlantic Ocean via the Bow River and Hudson Bay, the rest flowing to the Pacific Ocean via Wapta Lake, the Kicking Horse River and finally the Columbia River. At this time, Stephen was in its final season of being a turning point for steam helpers from Field and Lake Louise. Here, we drop off most of our train, leaving only a short string of empty flats between the diesel units and our caboose.

[This page] On a curved siding just beyond Stephen our westbound extra meets an eastbound extra led by another new General Motors set with FP7A No. 4060 on the lead. This engine spent its final years as VIA No. 1418. Spaced between a string of trailing outside braced boxcars is CPR G4aClass 4-6-2 No. 2701 (blt. CP Angus Shops, 10-19), being towed dead from assignment in B.C. to CP's Ogden Shops in Calgary for overhaul. This engine was scrapped in 1958.

[This page] Past Stephen we begin our descent towards the Spiral Tunnels and Field. The A-B-A diesel set is mainly coasting, especially with its light load of 5 flatcars and our caboose. As we approach Hector and Lake Wapta, in the distance, a Greyhound bus manages to sneak through on the Trans-Canada Highway. To the left are the Cathedral Crags, reaching over 10,000 feet, home of mountain goats and grizzly bears.

[Three photos] At Hector we go into the siding for a meet. Although by this time most trains were diesel-powered, some steam still stopped at this enclosed water tank, especially coming upgrade, the other way. In our case, the superior upgrade train is the second section of No. 2, the "Dominion," hauled by FP7A No. 4063. Its 15 cars include two diners, a "foreign" sleeper and an open-air "mountain observation car," from which a uniformed trainman observes us in passing. Note the little shingle-covered Hector station built right over the shore of Lake Wapta.

[Opposite] The 'Big Hill' is perhaps Canada's most spectacular piece of mainline railroading; at every turn there is another dramatic scene, causing film and time to be used up quickly. Going left into the bowels of Cathedral Mountain, we cross over our route twice in the Spiral Tunnels, then drop down into the Kicking Horse Canyon, going through several wood-framed snowsheds on the way. Our extra train terminates at Field, a crew change point, where the RPS gang lays over for a session of photography that we'll have to save for another book.

[This page]But I can't wait to show you this pair of scenes from the next day's trip, when the RPS crew continued west aboard the caboose of another extra freight. Down at Ottertail their train took the siding for No. 2, the eastbound 'Dominion.' Although by then officially dieselized, mechanical problems in Revelstoke that day resulted in the train being hauled by 'Selkirk' 2-10-4 No. 5920 (MLW 11-38; scr. 5-57). This is perhaps one of the last times that CP steam worked in mainline service through the Rockies. By the end of the year all steam (32 'Selkirks' and 40 others) was replaced by diesel (26 A-units, 20 B-units and 20 others).

CANADIAN CABOOSES

A Farewell Photo Tribute

Look through a typical handful of photos showing Canadian trains and you're likely to see no more than two or three cabooses. Usually, when a caboose was photographed it either showed up incidentally as the back part of a train, else it was the subject of singular ''roster shot.'' We've all heard how cabooses served as ''home away from home'' for train crews of the past, yet there is very little record of caboose life. Unfortunately we're now too late to rectify that, since many Canadian trains began to run without the familiar rear-end cars, starting in January 1990.

Canadian railways began to use rear-end crew cars by the late 1860's, usually little four-wheel ''bobbers,'' with the more common style appearing regularly by the 1880's. Though the term ''caboose'' was considered standard, eastern railroaders have often preferred to call their cars ''vans.'' While photos may be in short supply, written information about cabooses is even more scarce. Queries by this author to ''experts'' of both major Canadian railroads resulted in little more than shrugged shoulders. May some more peristant soul eventually present us with a written account detailing caboose developments, styles and assignments across the country.

From the Bytown Railway Society's **Canadian Trackside Guide** one learns the following statistics near the end of the Canadian caboose era:

1. In 1988 there were still some 1,710 cabooses listed on the rosters of Canadian railways.
2. Canadian National led with 904 of these, including 656 mainline steel models, 208 transfer and switching units, 16 wooden cars and 18 on the now-defunct Newfoundland narrow gauge.
3. Canadian Pacific was second, with 674 cabooses (down from 816 just ten years earlier), including 559 steel centre cupola, 68 steel end cupola and 49 wooden models.
4. As of 1988, there were known to be roughly 255 cabooses preserved in Canada. This number does not include most of those cars being used without wheels as cottages and sheds. Of these, 101 are former CN cars, 85 are from CP and 69 are from various other roads. The overall number should increase in the next few years, with retirement and disposal of most remaining cabooses from Canadian railways. Incidentally, only five pre-1900 cabooses are known to exist in the country, with four of these in Ontario and one in B.C.

[Below] It was March 16, 1952 when the photographer caught this Canadian National caboose-hop running from Fort Rouge to downtown Winnipeg. White-flagged Mikado S-2 No. 3584 (MLW, 1929) is trailed by wooden caboose No. 78808.

Lawrence Stuckey Photo/Paterson-George Collection.

[Below] A photographic mate to our colour front cover scene, this shot shows 2-10-0 No. 5784 (Blt. CP-Angus Shops 7-18; Scrapped 5-56) pushing on caboose No. 436911, which trails Train No. 978 headed east out of Minnedosa, Manitoba. According to the photographer's notes, Nos. 5782 and 2349 were on the head end and the date was April 19, 1953. He also notes that the marker lamps on trailing caboose No. 436503 (built in 1913) signify this to be Minnedosa's "East Pusher," which will later cut off the train but follow it beyond the 25 mile yard limit to Keyes, where the engine will be turned in order to assist a westbound freight back towards Minnedosa. The CPR's Minnedosa terminal was unique in that freight trains required helpers when leaving in either direction.

Lawrence Stuckey Photo/Paterson-George Collection.

Patterson-George Collection.

[Left] The marker lamps are still up on wooden caboose No. 436204 (Blt. 1911) as it is being brought from its train to the caboose servicing tracks in West St. John, N.B. by N2A 2-8-0 No. 3642 (MLW 3-11, scrapped 7-61). Note the leaf-springs on the van's archbar trucks, which were outlawed on interchange cars long before this September 1952 scene.

[Below] Flying the white flags of an extra, this branchline train passed through Forrest, Man. on July 18, 1958 with two cabooses and a string of flats. Spare tenders were sometimes carried in the West, especially in the late steam era, if water supply was limited or of poor quality along the way. D10 4-6-0 No. 909 (CLC 6-11) was scrapped less than a year after this photo.

Paterson-George Collection.

S. K. Bolton, Jr./RPS/AHW Colln.

[Left] This engine still survives! Passing the caboose tracks at Aroostook, N.B. on Aug. 4, 1956 is CPR No. 3388, one of five nearly-antique M3b 2-8-0's built by Schenectady in 1902 and assigned to this area until dieselization. Their relatively light weight allowed them to work branchlines off-limits to heavier power. No. 3388 can be seen today at the Canadian Railway Museum in Delson, Quebec. Beyond the engine is Aroostook's wooden station and freight house. The wooden caboose on the left is unusual for its door, made of narrow tongue and groove lumber. Caboose on right is steel-bodied.

[Below] In contrast to vintage No. 3388 is brand-new P2g No. 5405 (MLW 7-40; scr. 5-65), here leaving Outremont Yards, Montreal, for a break-in run on the Winchester Sub to Smiths Falls, July 21, 1940. Caboose behind the engine carried CPR officials and technicians from the Montreal Locomotive Works to monitor the new engine's performance.

Fred Sankoff/Paterson-George Colln.

John Poulsen Collection.

All three, CP Archives.

RPS Photo/AHW Colln.

[Opposite Page, Left] Among the earliest caboose styles were converted boxcars like C.P.R. "Conductor's Van" No. 3015 (from the 1880's) and C.P. 410398, photographed in Winnipeg at the surprisingly late date of September 1952. Both were originally all-wood boxcars with truss rods. Note No. 3015's interesting archbar trucks.

[Opposite Page, Right, Top] Some 351 numbers later on the CPR's caboose roster, three-windowed No. 3366 worked the Selkirk Mountains out of Revelstoke during the 1890's. Most later CPR cabooses were variations of this style.

[Center] One such variation was the side-door caboose seen here, also in the Selkirks. It was especially handy on mixed trains, where the side door was used for loading and unloading express or, in this case, snowfighting tools.

[Bottom] CPR "Standard Caboose" No. 436804 shows that at least up to the time of its 1922 construction most — or all — such vans were built with three windows, though they were later rebuilt for just two. Note that the cupola also had a third window, in the center, which was later covered, and that archbar trucks were used.

[This Page, Top]CPR No. 437267 was one of three all-steel bay-window cabooses built in 1948, mainly for transfer service. One of the three survives in a private collection near Calgary.

[Center] Leading candidate for "ugliest caboose in Canada" must be No. 437424, one of a fleet of 50 such units built by the CPR's Angus Shops in 1954. "Cabose Life on Company lines has taken on a new look," crowed the CPR *Spanner*, "with the accent on stepped-up efficiency, utility, comfort and a good measure of glamour as well." Among the features: all-steel construction, "semi-streamlined" center cupola, "and even a home-size icebox." At least one of them has survived.

[Bottom] No. 439707 illustrates the later, all-steel version of CPR's "standard caboose," seen here after roof walks and ladders had been removed.

John Poulsen Collection.

John Poulsen Collection.

Paterson-George Collection.

[Left] What's a passenger train doing with a caboose? A more careful look at this photo shows it to be an Express Train, without passenger cars or other conveyance for the crew to ride. The train is led by war-time Mikado No. 5423 (CLC - 5/43, scrapped 4/59), followed by two refrigerator cars and two baggage cars. Said retired CPR engineer Clayton Harris: "This was the Fruit Train from the Coast; it always had a caboose." Note that this one has a darker-than-usual paint scheme and lacks Canadian Pacific lettering, plus all but three numbers. Taken near the Spiral Tunnels on July 31, 1947.

Both, W. C. Whittaker

[Left] Taken on the same day as the above scene, this photo shows caboose No. 436722 (built 1922) on the rear of a long freight descending the Big Hill into Field, B.C., brakeshoes smoking down the line. A hot day and slow train caused the crew to open both cupola windows and back doors. Note the archbar trucks with leaf springs; also the straight cap-less chimney, a contrast to the elaborate H-shaped cap seen on the upper caboose.

[Right] This time we really do have a caboose on the back of a passenger train, which is No. 6, seen eastbound at Temple, B.C. in the Summer of 1952. A "Selkirk" is leading the six cars, trailed by caboose No. 436622 (blt. c. 1918). The explanation is simple: The caboose was assigned to a crew based in Calgary, which had brought a freight train to Field on the previous day. A shortage of passenger crews at Field meant they were called for an easy trip back home, having no choice but to bring their caboose along with them.

RPS Photo/AHW Colln.

W. C. Whittaker Photo.

[Right] CPR Pusher 2-8-2 No. 5812 (CP - Angus, 5/20, scr. 9/55) is seen at Field in 1951, pushing a CP reefer and towing Caboose No. 437113 (blt. 1944) and another.

All five, Clayton Harris Collection.

THE PLOW TRAIN

Railroading was tough in Southern Alberta during the first weeks of the 1950's. Trains ran late, or not at all, as many lines were closed by deep snow and high winds. The photos on this page were made by one of the brakemen of a plow train called to clear the line from Calgary south to Fort Macleod. At left, Engineer Clayton Harris and his fireman have just stepped from the cab of their D-10 4-6-0 to the top of a snowbank to have their portrait taken. Above them is a view of the train's caboose, CPR No. 436699 (built 1913) heading away through the snowdrifts. Notice the star-and-circle emblem above the cupola — some CPR conductors had a custom of marking their assigned cabooses with such symbols so they could spot them from a distance in large freight yards.

[Right] Trainman of CPR caboose No. 437013 (built c. 1943) eyes the photographer, while ahead of him is "Selkirk" 2-10-4 No. 5917, Canada's mightiest class of engine, named for the mountains up which it is pushing the freight train ahead.

[Below left] A red wooden caboose with markers brings up the rear of a westbound freight, in this well-known scene taken on Morant's Curve — named for the photographer — not far from Lake Louise.

[Below right] Another widely-published freight scene is this one, of a train crossing over itself in one of the famous Spiral Tunnels. Caboose 439871 represents the CPR's final development of such cars, with wide-vision center cupola, swivel seats and all-metal construction. Note the marker lamps, soon after replaced with electric lights.

H. W. Pontin/ RPS/AHW Colln.

Nicholas Morant/CP Rail.

Nicholas Morant/CP Rail.

Adolf Hungry Wolf Collection.

James A. Brown Photo

Paterson-George Collection.

[Upper left] Symbols of 1950's Canadian railroading include here: Semaphore signals; a clean wooden caboose (CNR No. 77106, built in the 1920's); an early-model diesel in fancy livery (in this case a CLC-built roadswitcher of unknown identity, wearing CN's green and gold); with a factory smokestack at work beyond the 40-foot steel boxcar. Taken in the Maritimes, June 1958.

[Upper right] International Caboose Mates — CN 76048 (blt. c. 1911) brings up the rear of a westbound Norfolk & Western freight, operating on trackage rights through southern Ontario over Canadian National's Cayuga Subdivision, near Cayuga, Ont., 15 July 1965. N&W's American-built four-window caboose No. 2607 appears well-kept, running on unusual trucks.

[Left] Wooden caboose 78077 (blt. c. 1930) has just come into Kingston, Ont. on a freight, while CN 2-8-0 No. 2566 is departing.

Lawrence Stuckey Photo/Paterson-George Collection.

Paterson-George Collection.

[Above] A CN caboose-hop enters Winnipeg on August 17, 1942 with S-2-c Mikado No. 3581 (MLW 1929) trailing wooden caboose No. 77854 (blt. 1920's).

[Right] A string of cabooses awaits assignment in CN's Lindsay, Ont. yard during the Fall of 1952. Several wayfreights — such as the one on the left — originated at Lindsay, so there were always vans on hand. Note that the second car has a CN-type body but CP-type cupola, darker than usual paint, and no lettering. Standard colour for CN cabooses was orange.

Paterson-George Collection.

Paterson-George Collection.

National Museum of Science & Technology

Paterson-George Collection

National Museum of Science & Technology

Bob Sandusky Photo.

[Opposite, top] Nova Scotia's Dominion Atlantic Railway ran as a subsidiary to the CPR, rostering some very distinctive equipment, including the four tail-end cars in these views. Oldest of these was Caboose-baggage car No. 98 (upper left), an all-wood car built in 1879. Details are not known for much-longer car No. 91 (upper right), seen at Yarmouth, N.S. in 1942. Car No. 92 (center right), seen at Windsor, N.S., was built in 1890.

[Opposite, bottom left] Cumberland Railway & Coal Co. caboose No. 501 is a two-tone van with interesting details that include truss rods, passenger-car trucks, and large front windows. Grand River Railway Caboose No. 8 was a 22-foot "shorty" van built by CP at its Angus Shops in 1923. Equipped with unusual curved archbar trucks, it is shown here behind the Grand River shops in July 1950, shortly before it was replaced by a longer, standard CPR caboose.

[Below, left] Temiskaming & Northern Ontario caboose No. 89 rides on passenger trucks in this 1941 view at North Bay, Ont. This was the line's last wooden caboose, the series having started with No. 50. Fading red paint shows the original coat of orange underneath. Note the slogan, "Ship and Travel by your Railway." T&NO steel cabooses with regular cupolas, such as No. 93 seen at North Bay in 1941, were numbered from 90 to 111 and painted silver with red ends. The oval "Safe Reliable Service" slogan remained even after the cabooses were relettered Ontario Northland.

[Bottom] Wooden caboose No. 65, in silver and red paint scheme, trails mixed train #209, leaving North Bay for Englehart on April 11, 1941 behind T&NO 2-8-0 No. 305.

All three, Fred Sankoff/Paterson-George Colln.

Bob Sandusky Photo.

Paterson-George Collection.

Paterson-George Collection.

[Top] Quebec shortline Alma & Jonquiere owned two diesels, four hoppers, a few old boxcars and wooden caboose No. 3007, when this photo was taken at Ile Maligne on Aug. 30, 1962. By then the 11.4 mile A&J no longer had passenger equipment, so this red caboose carried the occasional passenger between Saguenay Power and Alma. Colourful Alco-built switcher No. 102 was red, yellow and green.

[Below] This handsome 1890-built vehicle is one of the oldest cabooses left in Canada today. London & Port Stanley Railway had the C1 at London, Ont. when this c. 1952 photo was taken, a boxcab electric engine coupled to the far end. After a stint with the Ontario Rail Association, the van became a tourist information office for Tottenham, Ont., near the ORA's steam-powered South Simcoe Railway.

[Above] Ontario's Napierville Junction Ry. ran several American-built, 1912-era cabooses like this one, seen at Adirondack Jct. around 1952. Bought second-hand from the Delaware & Hudson RR, they had small, triangle-shaped bay windows instead of cupolas. Two of them ended up at the Canadian Railway Museum in Delson, Quebec.

[Opposite, top right] CN wooden caboose No. 78811 (built c. 1930) is seen here in the short-lived yellow livery and lettering of the subsidiary Great Slave Lake Ry., Canada's first line to reach the Northwest Territories. Roma, Alberta, 1965.

[Top left] Quebec Railway, Light & Power Co.'s caboose No. 3 is seen on the rear of an eastbound freight drawn by electric locomotive No. 33, stopped briefly at Montmorency Falls in May 1952. The stove is producing a good tail of smoke. QRL&P handled freight and passengers with electricity between Quebec City and St. Joachim, while CN operated with steam from there to Murray Bay.

Bob Sandusky Photo.

National Museum of Science & Technology.

[Bottom] Last Revenue Run on the Maritime Railway! It is Sept. 23, 1961 as the crew takes a final break around their caboose (built 1890's for the Intercolonial Ry., running on passenger trucks). Ahead is 2-6-0 No. 10 (bought from CN; built new for Temiscouata Ry.) while 2-6-0 No. 9 (also ex-CN) looks on.

Bob Sandusky Photo.

MIXED TRAIN CABOOSE RIDE

By Omer Lavallée

Back in the 1940's and 1950's, my annual holidays were eagerly awaited and carefully planned occasions, giving me a chance to get out of Montreal to go on long railway trips. At that time, in contrast with most amateurs who tended to be preoccupied with main-line railroading, I was very much interested in mixed trains and in branch line operations, feeling — correctly as it developed — that such services were the most vulnerable to highway competition and thus would be the first to go.

In 1953, I planned a trip to Western Canada and the United States with Ron Ritchie — another career CPR man — and we included a round trip out of Nelson, BC, which would enable us to ride the CPR's British Columbia Lake & River Service steamer "Minto", as well as the mixed trains on the Kaslo and Slocan Subdivisions. The services in the Slocan Mountains were geared to the "Minto's" leisurely schedule. Based in Nakusp on Mondays, spend the night at Robson West, return to Nakusp on Tuesdays, then make a return trip to Arrowhead and back on Wednesday. The same sequence would be repeated on Thursday, Friday and Saturday. Steamer and crew enjoyed a day off each Sunday at Nakusp.

On this occasion, on a Monday night in September 1953, we disembarked from Train No. 11, the "Kootenay Express", at Robson West. Passengers for the steamer had the option of spending the night on board — a fact not advertised in the public timetable — as there were no hotels nor settlement of any kind at Robson West. We were assigned a comfortable cabin, and, on Tuesday morning, enjoyed a full CPR-style dining-car breakfast.

The "Minto" cast off for the run to Kaslo promptly at 0530 (Pacific Standard Time). Being Canadian Pacific employees and naturally having made arrangement in advance, we spent most of the voyage in the wheelhouse with the "Minto's" captain, taking time off to have an excellent lunch before arrival at Nakusp at 1530. The steamer called at all of its eight scheduled ports of call, as well as most of the additional eleven flag stops in the course of the 153 km (95 mile) journey. Once again on Tuesday night, we were permitted to remain on board. We were up promptly at 0500 on Wednesday morning in order to have breakfast before boarding the mixed train for Rosebery which was due to leave at 0700.

Picking up our bags and bidding goodbye to the "Minto's" Oriental cook who had provided so liberally for our sustenance from his compact but complete galley, we made our way to the adjacent Nakusp station which its small railway yard to board mixed train No. 842 which, according to the timecard, would get us to Rosebery at 0850. From Rosebery, it was our intention to ride the train barge for 39 km (24 miles) down to Slocan City and there board another mixed train (also numbered 842) to return to Nelson in time to catch No. 11 that evening for Vancouver.

It should be borne in mind that, at that time, the steamer pier, station and yard at Nakusp were situated on a bench below the town proper. Subsequently, this facility was abandoned and covered by water as a result of the construction of the High Arrow Dam which raised the level of Upper Arrow Lake.

Standing in Nakusp yard under steam was Vaughan class M4g 2-8-0 No. 3480 (MLW 7-07), coupled to several freight cars and a caboose. There was no passenger car. Where, we wondered, was the mixed train? An inquiry in the station revealed that passengers for train No. 842 were so few and far-between that they didn't bother with the passenger car unless it was known in advance that there would be a large number of passengers. We were invited to board and ride in the caboose; we did so without hesitation and at the crew's suggestion, took seats in one half of the cupola.

The train left Nakusp on time and soon we were travelling along the valley with the Valhalla Mountains to the right, through Brouse, Summit Lake and Hills. The day was cool and overcast; not the best travelling weather. All too soon, we had covered the intervening 45 km (28 miles) and arrived at Rosebery where we were scheduled to transfer to the barge for the return to Nelson. By this time, however, the crew had become well acquainted with the fact that train-riding was a hobby, and it was suggested that we remain aboard for a further 55 km (34 miles) as far as Kaslo, in order to cover the whole Subdivision. Our desire to connect with No. 11 that night in Nelson was met by assurances that there was a bus departure from Kaslo which would get us into Nelson in plenty of time. Since passenger service had not been offered through to Kaslo since the 1920s, we needed no urging to remain on the train.

After the barge had arrived at Rosebery and No. 3480 had engaged in switching cars on and off the barge for Slocan City, we were off again. The scenery changed as the rails climbed grades of the order of 2% alongside Carpenter Creek through the deep precincts of Denver Canyon. Vestiges of the fabled Slocan silver-lead mining activity in the past came to mind at the ruins of Alamo Concentrator. Eventually, we arrived at Three Forks, 14 km (9 miles) east of Rosebery, where the train reversed at the lower leg of a switchback which was designed to allow the line to gain altitude quickly to continue the climb up Carpenter Creek to the summit at Zincton.

After backing its train for 1.6 km (1 mile), No. 3480 cleared the upper switch at Parapet, whereupon the train resumed its uphill, forward progress eastward. At this time, the tail track of the Parapet switch continued on as a spur for 5 km (3 miles) to the ghost town and mining camp of Sandon. This branch was noted for having the steepest main track grade on the CPR system: the 2 km (1 mile) nearest Sandon climbed at an average grade which exceeded 5%. There was still a small amount of concentrate being shipped from Sandon at that time and the maximum load for a class M4 2-8-0 on this spur was two empties up the grade and one load down. This was another "occasional" foray for train No. 842. However, unfortunately for us, there was no necessity for the train to visit Sandon on the day that we were riding.

From Parapet, the train continued around the base of Payne Bluff nearly 200 m (700 feet) below a ledge carrying the abandoned right-of-way of the original narrow-gauge Kaslo & Slocan Railway. Between Parapet and Zincton, 8 km (5 miles), the line climbed to 1,060 m (3,500 feet), the summit altitude for the whole Kaslo Subdivision. Along this section, the abandoned K&S roadbed could be clearly seen on the mountainside to the right. At one point, there was a deep lateral valley which was once spanned by a wooden trestle that remained for decades after the line was abandoned in 1909. On one trip early in the 1950s, the crew looked up to find that the structure had disappeared in the preceding week, a victim of old age and a heavy snowfall. At Zincton, there was some switching to be done at the Lucky Jim Mine, then No. 3480 resumed charge of its train for a leisurely run down grades ranging from 1.5 to 2%, for the rest of the distance to Kaslo.

As we approached Kootenay Lake, it was lunch time and our conductor became preoccupied with the preparation of the midday meal for the crew. The descent into Kaslo was by way of a switchback, the lower end of which went directly to the barge apron. Thus, it was necessary to remove the caboose from the rear end of the train. The train stopped with the caboose just upgrade from the switch to the lower line; the trainman set the van's handbrake then uncoupled it from the train, which moved up past the switch, then reversed to back down the lower line. As soon as the engine was clear and the switch reset for the main track, the conductor released the hand brake, then returned inside to attend to his cooking. As we were standing on a 2% grade, the car picked up speed rapidly, hurrying past the switch and around a curve on the tail track. An apprehension that we might be involved in an unplanned runaway was eased when we found that the tail track gradually steepened to about 5%, thus slowing down and eventually stopping the freewheeling van. While the conductor continued with his cooking, the car then began to roll in the reverse direction, gradually picking up momentum once again. As it approached the switch, we saw that the brakeman on the ground had reset it for the lower line. He swung on board as our speed continued to pick up, and brought the car to a smooth stop just a carlength from No. 3480 on Kaslo wharf.

The experience left us disinclined to accept a profferred invitation to lunch, as Ron and I were both feeling a little "green". Instead, we directed our steps towards the Kaslo bus terminal, to resume our interrupted trip to Nelson and Vancouver.

That part of the Kaslo Subdivision between Denver Canyon, its present (1987) eastern terminal and Kaslo, 49 km (30 miles), as well as the 5 km (3 miles) Sandon spur, was abandoned effective 15 February 1957.

[Below] Among the C.P.R.'s most unique operations were the combined train and boat services along the Arrow Lakes of eastern B.C. In this Summer 1947 view we see 4-6-0 No. 443 (Blt. CP-Angus 2-13, scr. 3-53) pushing its mixed train (caboose, coach, baggage car and two boxcars) up to the steamer "Minto," which has just arrived at the Arrowhead dock from Nakusp. After exchanging passengers and freight, the mixed train will head back to its mainline connection at Revelstoke. The "Minto" made her last run in 1954 and the rails to Revelstoke were taken up in 1969.

W. C. Whittaker Photo.

6203

5611

CN AT NIGHT

Hissing, softly chuffing, occasionally clanging to themselves — they filled the night air with power; the life of steam technology. Many have stood in awe beneath their presence — startled when pop valves lifted, letting steam spray noisily skyward. Many have trembled, eager to see the mighty iron breasts move — turning big wheels, making smaller parts go back and forth, up and down. Bells clanging to signal the move; dynamos whining, making rays of light shoot brightly through the blackness. A couple of toots on the whistle and it's off! Ah, to experience all **that** once again. . .

CN Steam at Night; multiple-flash time exposures shot at various locations in the East, circa 1950. The vintage prints lack data, but were taken by Dr. Philip Hastings and others of the Rail Photo Service for catalogue sale to rail-photo collectors around the world.

These prints courtesy Warren E. Miller, Railway Negative Exchange; Collection of Adolf Hungry Wolf.

ONTARIO WAYFREIGHT CP STEAM IN 1958

By James A. Brown

These three scenes capture a typical CP branchline freight in the twilight of steam operation. This is the Havelock-Peterboro-Lindsay-Bobcaygeon wayfreight on the 4th of September, 1958. A year earlier this train still carried passengers, as Mixed 605, daily except Sunday. On this day, its engine is D4g 4-6-0 No. 434 (blt. CP Angus 12-12), which was scrapped by following Spring. At left, approaching the station at Bethany, as seen from the fireman's window. Just over two miles from here is the junction with CP's Toronto-Smiths Falls-Montreal main line. Tracks seen here are now gone. Below, switching at Lindsay. Next page, the wayfreight running southbound along the Scugog River, entering Lindsay, Ont., as seen from cupola of its caboose.

Three Photos, James A. Brown

LAST STEAM REBUILDS AT CN STRATFORD SHOPS

By James A. Brown

It's the 29th of December, 1958 and the era of mainline steam is just about over in Canada. At Canadian National's motive power shop in Stratford, Ont., U1a 4-8-2 No. 6001 (CLC, 1923) has the distinction of being one of the last steam engines to receive a major overhaul. Here, it is being lowered back onto its wheels by a 200-ton Morgan crane.

Both pages, James A. Brown

995
6001
2721

Both pages, James A. Brown

[Opposite] Reassembly work is well under way on CN's N5b 2-8-0 No. 2721, following its overhaul and oil conversion at CN's Stratford shop, Dec. 29, 1958. Two pistons in the foreground await installation. Little more than a week later, No. 2721 is hostled past the shop by U2g 4-8-4 No. 6221. Both engines are receiving running tests following work done in the shops. Alas, in a few months both engines will be displaced by diesels. No. 6221 will be scrapped in March 1960, while 2721 will endure until October 1961. Meanwhile, S2a 2-8-2 No. 3552 receives its final paint detailing on Jan. 8, 1959, being the last steam engine to receive an overhaul at Stratford. Moved directly to storage, it was scrapped $1^{1}/_{2}$ years later without ever re-entering service.

[Above] Using the same tools and equipment as for overhauls, Stratford crews begin to scrap their own steam engines, as here with the remains of U1b 4-8-2 No. 6032 on Jan. 8, 1959.

[Right] In the bay adjacent to the scrapping stands 4-8-4 No. 6221, ready to head out for a few more months of work.

CANADIAN NATIONAL RAILWAYS

THE GREAT CANADIAN STEAM RAILWAY EXCURSION

It was **the** North American rail event of 1988 — three steam locomotives leading two different trains out of Vancouver on the same morning. Five full days of mainline steam action that included doubleheaders, mountain climbing and derailments. Photographic highlights follow on these next pages. (For the author's written report, see TRAINS Magazine, Feb. 1989.)

[Left] As part of this special event, former CPR 2-8-0 No. 3716 leads the first steam-powered freight over BC Rail tracks in decades as an Extra North here approaching the "Royal Hudson" tourist destination of Squamish, B.C. on this foggy morning of October 21.

[Below] After taking water, the steam freight waits on a yard track for BC Rail's daily RDC-powered Train No. 1 whose Conductor leans from the baggage door and exchanges momentary greetings during loading of passengers and mail.

Both pages, Adolf and Okan Hungry Wolf Photos.

[Above] While the freight extra heads north on BC Rail tracks, this passenger extra heads east over CN's mainline behind Canada's two largest active steam locomotives, CN 4-8-2 No. 6060 and CP 4-6-4 No. 2860, seen here at the Kamloops, B.C. depot.

[Above right] The passenger extra, crossing the Thompson River at Kamloops.

[Below right] It's been many years since Canada saw a steam-powered doubleheader running at mainline speeds. The train's 500 seats were sold out well in advance.

Both pages, Adolf and Okan Hungry Wolf Photos.

[Above] Late afternoon on the second day sees the excursion train arriving at CN's Blue River station, where the two engines will be watered.

[Below] It's well after midnight as the mighty 6060 makes a triumphant return "home" under steam to Jasper, Alberta, where the engine was on static display for ten years, starting in 1962. At the throttle is CN engineer Harry Home, who looked after the engine during its retirement years, dreaming of the day she would steam again. Now owned by the Province of Alberta, this magnificent "Mountain" class engine is ideal for hauling special-event trains through the Canadian Rockies.

[Above] The mountains of Jasper National Park make an ideal backdrop for scenes of this historic doubleheader on the third morning of the great excursion. Plans had been for the two engines to lead the train west to Prince George, but fast running to make up lost time the previous night left No. 6060 with a loose rod that forced the engine to remain behind in Jasper, after these dramatic photos were taken.

[Below] Not since the late 1950's has CN's Jasper yard seen three steam engines, as the doubleheader rolls slowly past 4-8-2 No. 6015, which replaced No. 6060 on display here in 1972. The excursion train's 15 cars included maroon ex-CP stock from BC Rail, along with ex-CN cars in VIA Rail's blue and yellow. A sixteenth car was lost on the first day, after it derailed at speed not far out of Vancouver. Fortunately, there were no injuries in this mishap, though a second derailment occurred the next morning when parts of both engines left the tracks in CN's Kamloops yard.

[Above] It is unlikely that CN's Jasper roundhouse will ever see two live steam engines again, as it did during the early morning hours of October 23, 1988. Plates above the engine pilots advertise their respective provincial homes and owners.

[Right] Running alone instead of doubleheaded, British Columbia's famed "Royal Hudson" leads the 15-car excursion train westward over the Rockies, through Yellowhead Pass.

[Below] A hot axle bearing causes more problems, as the "Royal Hudson" pauses with its train at Redpass Junction, where CN's Vancouver and Prince George lines diverge. The train finally reached its destination some 10 hours late.

Three photos, Adolf and Okan Hungry Wolf.